Index

Foreword

I.   Introduction
      A.   Archimedes' Lever
      B.   The Antikythera Mechanism
      C.   Thesis

II.   Mechanology
      A.   Protologism
      B.   Oxford English Dictionary Definition
      C.   Biology, History, Archaeology, Engineering
      D.   Glossary

III. The Four Different Types of Machines
      A.   Simple Machines
      B.   Compound Machines
      C.   Complex Machines
      D.   Autonomous Machines

IV. Methods of Modification
      A.   Variation
      B.   Consolidation
      C.   Integration
      D.   Convergence

V.   Object(s) of Work
      A.   Function
      B.   Task
      C.   Product
      D.   Utility

VI. The Demarcation Problem

VII. Conclusion

VIII. Citations

Publication Date: 7/11/2017
ISBN/SKU: 9780692920725
SC: COM080000
SC2: TEC000000

Foreword

"...we find ourselves almost awestruck at the vast development of the mechanical world, at the gigantic strides with which it has advanced in comparison with the slow progress of the animal and vegetable kingdom...

Take the watch for instance. Examine the beautiful structure of the little animal, watch the intelligent play of the minute members which compose it; yet this little creature is but a development of the cumbrous clocks of the thirteenth century— it is no deterioration from them. The day may come when clocks, which certainly at the present day are not diminishing in bulk, may be entirely superseded by the universal use of watches, in which case clocks will become extinct like the earlier saurians, while the watch (whose tendency has for some years been rather to decrease in size than the contrary) will remain the only existing type of an extinct race.

We regret deeply that our knowledge both of natural history and of machinery is too small to enable us to undertake the gigantic task of classifying machines into the genera and sub-genera, species, varieties and sub-varieties..."

"Darwin Among the Machines"
Cellarius (q.v.), 1863

In 1859, Charles Darwin published *The Origin of Species*. The academic climate across the globe was filled with his students and critics. Four years later, Cellarius applied the revolutionary ideas of selection and change to invention and technology in his essay, "Darwin Among the Machines". In it, he claimed that machines were a kind of mechanical life undergoing constant evolution. What is intriguing in Cellarius' writing above is not just his claim that machines are changing via a consolidation of function, but the motion towards a method of classification; the genera and sub-genera.

From the writings of Thomas Hobbes to John von Neumann, thinkers have studied machines and their role in our civilization. Yet, Cellarius was one of the first known to contemplate the existence of machines as an object of study by observing the earliest processes of consolidation and miniaturization.

Further still, the most revealing part is not just the writing but how it was published. Titled after Darwin, Cellarius' essay was released in an obscure newspaper in Christchurch, New Zealand, one of the furthest reaches of the world for industrial Europe. Additionally, "Cellarius" was a pseudonym used to mask the identity of the true author, known in England as Samuel Butler.

It is possible that author Samuel Butler was tentative of his own ideas, uncertain of their substance, or concerned with their reception. Nonetheless, to claim in the 19th Century that machines were changing and that machines should have their own system of taxonomy and classification was thought-crime. It was an idea that would have one's past and future work discredited.

Samuel Butler was both a critic and a student of Darwin, both serious in his ambition and clandestine in his satire. He was one of the first and only to describe machines as worthy of our deepest analysis, to claim we are in the midst of an experiment that echoes the prehistory of human intelligence and the origins of life.

Fast forward a century to the digital universe and Butler's prospection of our computer programs and worldwide networks combining to produce a new evolutionary theater is all too vivid.

I. Introduction

The ancient world was fascinated by machines. The craft of invention preoccupied some of the greatest thinkers and inventors from Archimedes to Heron of Alexandria. One very old adage attributed to the great mathematician Archimedes goes, "Give me a place to stand and with a lever I shall move the whole world" ($\delta \tilde{\omega}\varsigma \mu o \iota \pi \tilde{\alpha} \sigma \tau \tilde{\omega} \kappa \alpha \grave{\iota} \tau \grave{\alpha} \nu \gamma \tilde{\alpha} \nu \kappa \iota \nu \acute{\alpha} \sigma \omega$).

Archimedes didn't invent the lever itself but he was fascinated by its principles, claiming that it is one of the most basic yet effective machines. The Ancient Greeks called these mechanisms simple machines. Today we call them tools.

The idea of a simple machine originated with Archimedes around the 3rd century BC, who studied what is known as the Archimedean simple machines: lever, pulley, and screw. Later, Greek inventor Heron of Alexandria in his work *Mechanics* defined the classic five simple machines (excluding the inclined plane). He listed the five basic devices that could "set a load in motion"; lever, windlass, pulley, wedge, and screw. However, the Greeks were limited by the lack of the study of dynamics; the tradeoff between force and distance and the concept of work.

In 1586 Flemish engineer Simon Steven introduced the mechanical advantage of the inclined plane and it was included with the other simple machines. The complete dynamic theory of simple machines was worked out by Italian Scientist Galileo in 1600 in *Le Meccaniche* (On Mechanics), in which he showed the underlying mathematical similarity of the machines. He was the first to understand that simple machines do not create energy, only transform it. Today, they can be defined as the six simplest mechanisms that use mechanical advantage (also called leverage) to transform force.

What makes Archimedes' adage profound today is not just his expression that there is no limit to the amount of force amplification that can be achieved by using mechanical advantage. It is the concept that the lever is one machine amongst others from which all machines derive. It creates a way to categorize machines based on evidence. From this framework, we continue to develop a systematic method to study new, more complex machines with evidence.

One very interesting machine from the ancient world is the Antikythera Mechanism. A diagram of it is seen on the cover. It was found one hundred feet beneath the seafloor after 2,000 years of resting under the ocean water. It is considered the world's first analog computer.

Resembling a mantel clock in appearance, this machine and its parts display the celestial motion of each of the five planets visible to the naked eye- Mercury, Venus, Mars, Jupiter and Saturn. Whereas Archimedes claimed to move the world with a lever, the Antikythera Mechanism so too moved the planets on a dial and displayed the phases of the moon and the timing of the lunar and solar eclipses.

Obscured by corrosion of rusted bronze are traces of technology that resemble modern machines; gears with neat triangular teeth, interlocking wheels, and interrelated, moving parts. These simpler parts are the simple machines- the levers, wedges, and screws from which all more complex machines derive. Hidden in the discovery of the ancient astronomical computer at the bottom of the ocean is an utterly modern mechanism. What was found off the coast of the Antikythera islands in Greece in the 20th century was the single greatest craft of invention among the simple machines of the ancient world.

In this work, we make these distinctions by introducing the examination of machines as an object of study, or "Mechanology." Consequently, we will (1) explain the need to study machines in a gestalt way, (2) we will clarify and improve upon the definition of what a machine is, (3) define what the four different types of machines are and how they can each be classified by different pieces of evidence, and (4) ultimately, suggest there is an evidence based way in which they are modified.

II. Mechanology

Fusing the Doric word μα χ α ν ά (*makhana*), and -ology, 'study of', we'll use the new term **Mechanology** for the *study of machines*. Defined, Mechanology is the branch of knowledge that deals with the study of machines.

Mechanology can be described as a protologism[1] for a social science discipline concerned with the analysis of technological artifacts, including their distinction from each other by their composition, utility, variation, and classification. In general, Mechanology recognizes the *four* distinct definitions of machine, the simple machine as the basic unit of all mechanisms, and convergence as the engine of modification of mechanisms over time.

---

[1] Moore, Andrew (January 2011). "The hypothesis' ambassador". BioEssays. 33 (1): 1. doi:10.1002/bies.201090064. Recognising the preliminary (or even want-to-be) nature of many protologisms, Mikhail N. Epstein the American literary theorist and thinker coined his own: 'protologism', which refers to a neologism that is introduced when an individual or individuals find that a specific notion is lacking a term in a language, or when the existing vocabulary is insufficiently detailed. The law, governmental bodies, and technology have a relatively high frequency of acquiring neologisms and protologisms.

Today, when we think of machines, we think of power tools or complex devices. The definition according to the Oxford English Dictionary:

"Ma·chine[2], məˈSHēn/, **noun**"

- An apparatus using mechanical power and having several parts, each with a definite function and together performing a particular task.
- *i.g. 'a fax machine'*
    1. modifier: A coin-operated dispenser.
    2. i.g. 'a cigarette machine'
    3. technical: Any device that transmits a force or directs its application.

The definition above is but one of our generally agreed upon modern explanations of what a machine is. A current consensus is that:

1) A device that must have several parts.
2) A device that must use mechanical power.
3) A device that must have a definite function to perform a task.

The aim of the following chapters is to determine the conditions which are common to all

---

[2] Shoberg, Lore. *Machine*. New York: McGraw-Hill, 1973. Print.

machines in order to decide what it is, among its great variety of forms, that essentially constitutes a machine. The study of the constitution of machines naturally divides itself into two parts, the one comprehending the classification of the great variety of machines, and the other, the methods by which machines are modified over time. This work introduces the concept of a field of study committed to problems of classification and variation.

Consider, what is the singular most biologically-significant factor to humanity's survival? As recent philosophers and thinkers have written, the answer may be our technology. Technology is arguably the predominant reason that we live safer, longer, and healthier than ever before, particularly when we include medical technology, sanitation, antibiotics, vaccines, and communication technologies. Complex systems of machines such as satellites, smartphones, and the internet are all contributing to the pace and complexity of technological innovation at an exponential rate. What, however, qualifies as technology, or as a machine? Is it the telegraph or the phone, the typewriter and the computer, the bicycle and the car? What is the principle that these things all share? What precisely is the definition of a machine? These questions are something that a study of machines will help us qualify with evidence.

We have the word technology and the discipline of engineering as adequate descriptions for all machines of all different purposes and complexities. However, technology and engineering are the applications of scientific and mathematical knowledge to practical purposes, for example "...advances in computer technology.[3]" Technology and engineering are the applications of knowledge in which we create machines; the collection of techniques and methods used in the production of goods and services. Technology provides us with a broad variety of machines, from submarines to thinking machines. Argued in this work, is that there should be a method to study and classify machines.

The study of machines is part the biological-the collection of specimens and their classification. It is part the historical- the investigation of the past. It is part the Archaeological- the unearthing of ancient technological culture. And it is part Engineering- the cleverness of invention.

At present, there are vague and indistinct methodologies, divided by a lack of a gestalt and coherent discipline. In attempting to place the theory of the constitution of machines upon a new basis we will clarify the umbrella term 'machine' and provide a new nomenclature. This subject has hitherto been

---

[3]Shoberg, Lore. *Technology.* New York: McGraw-Hill, 1973. Print.

treated unscientifically. Namely, after acquainted with a new invention, we leap entirely over its components and processes which have furnished the results and regard it with only questions of use.

Mechanology considers machines as divided into four distinct categories:

i. **simple machines** or tools: a device used to apply a force to carry out a particular function.
ii. **compound machines**: a mechanism of two or more different simple machines applying mechanical power having several or more interrelated parts for the accomplishment of a certain task.
iii. **complex machines**: an apparatus of two or more different compound machines using or applying power sources - such as mechanical or electrical power - that is capable of producing both tasks and functions.
iv. **autonomous machines** (artificial intellects, 'artilects', or intelligent machines): a network of two or more complex machines measured by the expenditure of power sources, with the capability of utilizing and creating machines towards optimizing its own goals.

Mechanology considers machines as modified over time through four distinct methods:

i. **variation**: a change in feature but not function, i.e. size, shape, and color.

ii. **consolidation**: the change in the number of different functions.

iii. **integration**: the change in production via numerous functions and tasks.

iv. **convergence**: the change in the connected state of a machine, i.e. the tendency of all machines to reach a connected state.

Mechanology considers the types of machines as defined by four objects of work:

i. **function**: work that is done by a simple machine without the use of an external energy source.

ii. **task**: work that is done by compound machines with the consolidation of mechanical energy or mechanical advantage.

iii. **product**: work that is done by complex machines with the use of external and/or mechanical energy to integrate tasks and functions.

iv. **utility**: work that is converged by the autonomous machine itself to optimize goals requiring the use or indistinguishability between machine and energy.

Complexity of Machine	Method of Modification	Object of Work	Transformat-ion of Energy
Simple	Variation	Function	Applied by hand
Compound	Consolidation	Task	Mechanical Advantage
Complex	Integration	Product	Types of Energy
Autonomous	Convergence	Utility	Quantity of Energy

III. The Four Different Types of Machines

Simple Machines

_____

The six **simple machines** are an essential part of modern physics and were established early by the ancient Greeks. They are the wedge, lever, inclined plane, screw, wheel and axle, and pulley.

An analogy is that in Biology there are twenty amino acids, the building blocks of all biological life. An imperfect way to regard simple machines is as the building blocks for all other machines.

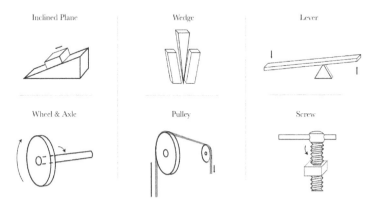

Inclined Plane   Wedge   Lever

Wheel & Axle   Pulley   Screw

What is a simple machine we all know? In fact, a hammer is a lever- a simple machine- a lever with a heavy end on it used to carry out the particular function of, well, hammering nails. Indeed, tools are frequently simple machines or composed of them and can occasionally be understood as equivalent but not synonymous. They simply *vary* in size, shape, and color, in that they change via variation.

So, in fact, Archimedes' lever and all levers are simple machines. Defined: a tool used to apply force to carry out a particular function.

We have the origins of machines, the building blocks, but what next?

Compound Machines

———————————

The next classification are pairs, also known as kinematic pairs, or **compound machines**. By this we mean two or more different simple machines working together, interacting via motion, contact, or mechanical arrangement. These are the machines we speak of: pencil sharpener, scissors, wine bottle opener, wheelbarrow, bicycle, stapler, the list goes on. We don't typically think of them as machines, but they are. Most of the machines in the world are compound machines.

With a compound machine, a smaller amount of force can be used to move an object, a product of all the simple machines working together[4]. This framework was pioneered by Franz Reuleaux in Berlin in the 19th century, where his seminal paper, *Outlines of a Theory of Machines*, introduced the principles of mechanical engineering that differentiated between simple and compound machines.

Reuleaux wrote there are lower pairs and higher pairs, turning pairs, screw pairs, cylindrical

———————————

[4] Reuleaux, Franz, and Alex B. W. Kennedy. *The Kinematics of Machinery: Outlines of a Theory of Machines (1876)*. Whitefish, MT: Kessinger Publishing, 2010.

pairs, spherical pairs, sliding pairs, and rolling pairs. An example of a sliding pair is a 'pair' of scissors and a rolling pair would be a wheel rolling on a flat surface. What does this mean, broken down?

Scissors are composed of two levers and two wedges which rotate on a fulcrum in order to produce a cutting motion. Consider the image below in what makes it a compound machine, can you identify items A through E?

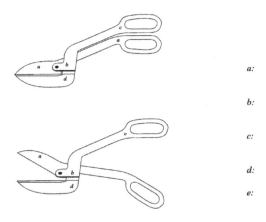

a:

b:

c:

d:

e:

With two or more simple machines interacting via motion, a pair of scissors can be described as a compound machine. Defined fully, a compound machine is a mechanism consisting of *two or more different* simple machines using or applying mechanical power having several or more *interrelated* parts or kinematic pairs for the accomplishment of a certain task. Items A through E would be in order: wedge, fulcrum, lever, wedge, lever.

When taken apart, scissors are composed of two levers and two wedges which rotate on a fulcrum in order to produce a cutting motion. With two or more simple machines interacting via motion, a pair of scissors can be described as a compound sliding-pair machine.

Similarly, a bicycle is a compound machine that is made up of pulleys and wheel and axle. When disassembled, two or more simple machines are interacting via motion of a rolling pair; the wheels rolling on the surface. Defined, a bicycle would be a compound rolling-pair machine.

Furthermore, a common compound machine is a cylinder and piston, frequently found as a component in the engine of a car. A cylinder and piston is considered a lower pair, a compound machine with a link and pair by which the surface has contact between them. Described, a cylinder and piston would be a compound higher-pair machine. If the engine of a car often has four or more pistons, what type of machine is an engine? Moreover, what type of machine would a whole modern car be with so many numerous machines working in concert?

Complex Machines

So we have amino acids and simple machines, the basic building blocks. We have compound machines composed of two or more simple machines not unlike RNA & DNA is composed of two or more different amino acids. But what is the next phase of the classification of machines? An analogy from Biology may be the equivalent of a prokaryote.

A **complex machine** is a mechanism with *two or more different compound machines.* The origins of complex machines first emerged in the industrial revolution of 18th and 19th century England- the time of Samuel Butler and Charles Darwin.

An analogy is that Biologists classify Prokaryotes for their varied source of energy inputs from photosynthesis to organic compounds. Mechanical, Electric, Hydro, Magnetic, Gravitational, Chemical, Nuclear, Radiant, Molecular, and Thermal are the classifications of a complex machine, also similarly classified by their energy inputs. A complete definition for a complex machine is a mechanism of two or more different compound machines using or applying power sources such as mechanical or electrical power capable of producing both tasks and functions.

What is a good example of this category? Is it the kind that Butler would believe have take over as the dominant species of Earth? So far, no. Consider a simple example, a projector from a movie theater, a complex machine. When disassembled, projectors are composed of many different simple and compound machines, responsible for transporting and carrying the slides of film in front of a bulb, powered by electricity, which projects the image onto a screen.

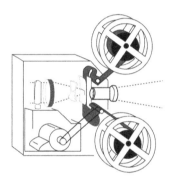

This machine is one that involves numerous other machines and is also requiring a unique energy source- electrical power- to operate: a *complex* electro-mechanical machine. The interaction of these two or more compound machines within the projector uses electrical power and produces the end result of projection- exemplifying the definition of a complex machine.

Let's consider the other types of energy that complex machines can utilize. Consider a steam turbine, or perhaps one of the very first steam turbines, invented only a couple hundred years after the Antikythera Mechanism in ancient Greece. This device was invented by Hero of Alexandria and is the first recorded example of a steam engine.

When a central water container is heated, torque is produced by steam jets exiting the turbine. Using numerous compound machines and its own, unique source of power, the Aeolipile could be described as a complex thermal-mechanical machine.

However, what's another example of an ancient complex machine? One of the oldest complex machines is the watermill. Using moving water as it's power source, numerous compound machines generate mechanical processes to produce the end product of grinding, rolling, or hammering. Thus, a watermill could be defined as a complex hydro-mechanical machine.

We have studied the origins of machines, but what of their pinnacle? In Biology, there is the Eukaryote, from which all complex life as we know it has arose.

Autonomous Machines

The final classification that a machine may fall under is **autonomous**. An analogy borrowed from Biology is that a Eukaryote is built from millions of Amino Acids, the building blocks of life. So too is an autonomous machine the result of a long process from the millions of building blocks of simpler machines. Truly autonomous machines, thinking machines, are yet unrealized but we extend a classification to them by all the methodologies applied to other classifications of machines: energy, goal, complexity, and method of change.

A note to make is that we are discussing the physical nature of an autonomous machine; its tangible self. One way to consider this perspective is that true AI is an emergent property of an autonomous convergent network of machines. The analogy is that a software program is a set of instructions that give emergence from the hardware of a computer, the same in that the consciousness is an emergent property from the hardware of the brain.

In AI Theory, there are a few pre-existing delineations that can be useful to a classification of autonomous machines. The first is the description of *reactive machines*[5] much like Deep Blue. Deep Blue was a chess playing computer developed by IBM in the mid 1990's, able to calculate millions of moves a second and make the most optimal move amongst the possibilities. However, Deep Blue can not form memories or use past experiences to inform decisions. It is purely a reactive machine. Furthermore, *limited memory machines* are machines that can look into the past. Self driving cars can do this by observing another car's speed and direction and identifying specific objects and monitoring them over time. Finally, *theory of mind machines* and *self-awareness machines* are machines that can form representations about themselves and others and have 'consciousness'. In this description, reactive machines and limited memory machines are not autonomous and can both still be classified as Complex Machines. Specifically, a supercomputer is computationally necessary for an autonomous machine, but having a supercomputer does not predicate an autonomous machine itself. Consequently, Deep Blue is not an autonomous machine.

Another classification in AI theory is Artificial Narrow Intelligence (such as Deep Blue), Artificial General Intelligence (such as the general intelligence

---

[5] Arend Hintze, The 4 Types of AI, Michigan State University, 2016

of a human), and Artificial Superintelligence[6] (the intelligence of all humanity and above). This method while useful, is value-laden and is a qualification that relies upon a general definition of human intelligence as opposed to a physical classification.

Whereas complex machines are defined with evidence by the *types* of power sources they rely upon, an autonomous machine can only be defined with evidence by the *quantity* of energy it consumes. It is of no import to an autonomous machine how many different types of energy sources it employs, but rather, simply how much and on how large a scale. Further, it is of no importance to a Mechanologist when classifying machines in the types of goals an autonomous machine may have, but simply, *how much* energy it utilizes for optimization of those goals. Measuring its usage of energy is the evidence needed to quantify and classify it in comparison to other complex or compound machines.

How do we measure the quantity of energy that an autonomous machine, an artificial intellect, utilizes? The answer is that a system of benchmarks must be applied in order to provide a common nomenclature in the same regard as the six simple machines, kinematic pairs of compound machines,

---

[6] (Kurzweil 2005, p. 260) or see Advanced Human Intelligence where he defines strong AI as "machine intelligence with the full range of human intelligence."

and akin to the varying energy source of complex machines. This system of benchmarks for autonomous machines is the largest hypothetical degree of energy consumption we know of, also known as the Kardashev Scale[7]. It is most often used, interestingly, to hypothesize about the level of non-terrestrial technological advancement, e.g., if a hypothetically non-terrestrial civilization utilizes all the energy of its local star for its technology. It contains three 'types' measured by energy consumption.

Applied, a type 1 autonomous machine relies upon local power sources and is contained in its local solar system. It uses or stores energy which reaches its location from a neighboring star.

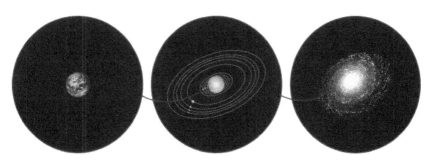

Type I : 10^{16} W   Type II : 10^{26} W   Type III : 10^{36} W

To compare on our terms, Earth civilization is still a 0.5 on the Kardashev scale. It doesn't wholly

---

[7] Kardashev, Nikolai (1964). "Transmission of Information by Extraterrestrial Civilizations". Soviet Astronomy

29

register. Human technology still utilizes the things it finds underground to burn as energy; fossil fuels.

For the available energy in the Earth-Sun system, this type 1 quantity is between 10^{16} and 10^{17} watts. A type 2 autonomous machine harnesses all the energy of a star in its own solar system. This means it is capable of utilizing or channeling all of the radiation output of its star- in the Earth-Sun system, this would be 4×10^{26} watts.

A type 3 autonomous machine consumes the energy on a scale comparative to a host galaxy. This means that it is in possession of energy on the scale of its own galaxy. In the Earth-Sun-Milky Way system, this would be 4×10^{37} watts. Consequently, a complete definition of an autonomous machine, artificial intellect, or intelligent machine is a network of networks between two or more different complex machines capable of utilizing and creating machines with the capacity to optimize its own goals, measured by the amount of power consumed, type I, II, & III.

On The Classification of Machines

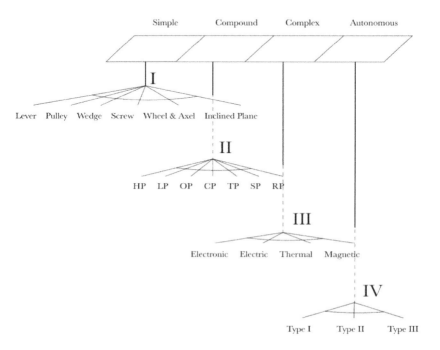

IV. Methods of Modification

    With a system of taxonomy and classification in place, the second largest problem facing a study of machines is a methodology in which to examine the modifications that can occur.

    The primary method of modification that describes how simple machines change is called variation. When a lever is used as a crowbar to open a door it is still defined as a simple machine, a lever. Indeed, when a baseball bat is used as a tool in recreation to hit a baseball, it is still a simple machine, a lever. Consequently, simple machines can variate by size, shape, color, and or in combination with function. To complete the definition, variation can be described as a change in features but not function, and in this regard, variation is how simple machines change.

    The second type of change describes the changes of compound machines and is called consolidation. When a pair of scissors is used, it employs four total simple machines; two levers and two wedges. We can describe the change in the machine from simple to compound as having consolidated the functions of the lever and the wedge to produce the task of cutting. A further example; a bicycle is composed of two wheels and axles, numerous pulleys and levers. When it is cycled, the consolidation of simple machines that produce a single function can be described with a compound machine (bicycle) that changes the functions to produce a function of cycling. Each of these examples are compound machines because they involve two or more simple machines and require mechanical energy.

The third type of change describes the changes of complex machines and is called integration. When a car is driven, it employs numerous simple and compound machines such as wheel and axles, levers, wedges, and pulleys, all which make up the frame, wheels, and engine as well as additional components such as lights, displays, circuit boards, and computers. When it is driven and the interrelated parts are applying and or using numerous sources of power they have integrated the functions, tasks, and production of simple and compound machines into a complex machine in order to 'drive'. A century ago, the most common type of machine was likely a compound machine; only rarely would you find light bulbs, telephones, and cars- machines requiring no kinetic or mechanical energy instead of the more complex electrical, chemical, and thermal energy to work. Rarer still, would you find light bulbs, telephones, and cars *all* in the same machine. A full definition can be that complex machines change by integrating with other machines, moving parts, and power sources to create a product.

One note to make when examining modification is that two machines are considered one if they share power source(s). An apparatus of different complex and simple machines are a gestalt if they rely on one collectively shared source of energy. A machine can not be considered integrated if it has its own separate or segregated power.

The last type of change describes the changes of autonomous machines and is called convergence. When a sufficiently advanced autonomous machine can optimize its own goals it can utilize and or create other machines. Machines can converge in that if they are sophisticated enough, they can choose how to change. A complete definition may be that convergence is a tendency of autonomous machines to acquire the function, tasks, production, and utility of other simple, compound, and complex machines-for all machines to eventually reach a connected state. This type of change is unique in that simple, compound, or complex machines can not change on their own, yet autonomous machines can. A compound machine is changed vertically and laterally, in that it can integrate into a complex machine or that it's constituent parts of simpler machines can variate subtly, yet a sufficiently advanced autonomous machine can initiate change of its own component parts and complexity.

V. Object(s) of Work

To provide a classification of machines and the mechanisms by which they change, a third definition must be provisioned. This is a definition of work, or its ultimate form, utility. The first means of work is considered a function, completed by simple machines. Consider crowbars, baseball bats, brooms, seesaws, and bottle openers are all different tools, yet the same simple machine. They variate by weight, size, shape, and color, and most significantly, by the function that they complete. The first definition of work used to describe machines is a function, defined as work that is done by simple machines without the use of external energy sources.

The next example of work completed applies to compound machines and is considered a task. Consider a pencil sharpener, typewriter, and bicycle. These are all different compound machines yet they all complete a task no matter the variation amongst their constituent parts (changes in shape and size), or the consolidation of their different simple and compound machines. They don't require power sources, but rather rely on human input to cycle, human input to sharpen a pencil, and human input to type upon a typewriter. Consequently, the work that is considered a task can be defined as work that is done by compound machines with the use of mechanical energy or mechanical advantage.

The essential dichotomy in the utility of machines is that simple and compound machines require kinetic or mechanical energy to operate, in that there is a force applied to them. However, complex and autonomous machines differ because they do not require a specific force applied to operate. They run on different types and amounts of energy sources.

The third means of work is considered a product, describing numerous complex machines integrated with other complex, compound, and simple machines. A product or production differs from tasks in the variety of tasks it can complete through its reliance on the use of external and or mechanical energy. Consider a car and a smart phone, both are capable of numerous functions, tasks, and products and require an energy source. A complete definition can be described as work that is done by complex machines with the use of external and mechanical energy to produce tasks and functions.

Finally, the fourth means of work is considered utility, describing the completion of product, task, and function by autonomous machines. An autonomous machine that converges upon any additional machine can be said to create utility, in that it is work that is determined by the autonomous machine itself to optimize goals requiring the use or indistinguishability between machine and energy.

VI. The Demarcation Problem

—————————

Analysis of what constitutes a machine presents us with the dual problems of how we classify them and how they change. Indeed, machines may be classified by simple, compound, complex, and autonomous, each growing in complexity, utility, and measured by the way in which they transform energy. Further, machines may be increasingly understood by the principles of modification which dictate how they change; variation, consolidation, integration, and convergence. Ultimately, these concepts aim to fill a gap in our lexicon, between technology and machine, to shed light on a field of study, and to provide analysis into simultaneous problems of classification and variation.

However, objections must be raised such as, "what is the point of a study of machines, why study them, and isn't technology suitable enough?"

A study of machines may be useful to many branches of knowledge. In fact, we ought study them not only out of a passion for studying and understanding machines but rather the lingering question, "why isn't technology as a word suitable enough?" Technology is the application of scientific knowledge to industrial means, not the study of its results nor the classification of its artifacts.

Engineering can assist in distinguishing from different machines. Engineering is essential evidence to Mechanology, but the essence of engineering is invention and creation of things not in nature rather than the classification of existing mechanisms. *Fung et al*, in the classic Engineering text, *Foundations of Solid Mechanics*, elucidates the context of Engineering and Science:

> "Engineering is quite different from science. Scientists try to understand nature. Engineers try to make things that do not exist in nature. Engineers stress innovation and invention. To embody an invention the engineer must put his idea in concrete terms, and design something that people can use. That something can be a complex system, device, a gadget, a material, a method, a computing program, an innovative experiment, a new solution to a problem, or an improvement on what already exists. Since a design has to be realistic and functional, it must have its geometry, dimensions, and characteristics data defined. In the past engineers working on new designs found that they did not have all the required information to make design decisions. Most often, they were limited by insufficient scientific knowledge. Thus they studied mathematics, physics, chemistry, biology and mechanics. Often they had to add to the sciences relevant to their profession. Thus engineering sciences were born."[8]

---

[8] *Classical and Computational Solid Mechanics, Y C Fung and P. Tong.* World Scientific. 2001.

Without the classification of all machines on a spectrum, we are unable to distinguish between the varying degrees of complexity with evidence and falsifiability. We are left with the definition found in dictionaries, "An apparatus using mechanical power and having several parts, each with a definite function and together performing a particular task".
Engineering is not a science, but rather a collection of sciences for a design or application- an invention. Mechanology then, is the study of machines to define and classify using the principles of engineering and physics.

A similar case study is Robotics; a practice of and assembly of many different machines. However, Robotics is an engineering discipline, whereas Mechanology is a social science discipline. Robotics is an active practice inside of the field of engineering, as opposed to a classification and study within the field of social science.

Furthermore, consider the study of Artificial Life, otherwise known as ALife. As currently defined, Artificial Life is the study of fundamental processes of living systems in artificial environments. It is not the use of evidence to delineate between mechanical devices and technological artifacts.

With so many branches of knowledge all around us, from the study of beauty in aesthetics, to the study of music in musicology, to the study of volcanoes in volcanology, to the study of stamps in philately, the study of flags in Vexillology, and the study of the future(s) in Futurology, ought we not study machines? It may be that a more appropriate terminology for the study of machines is required, or even that the idea itself evolves irreconcilably with the original ideas presented- but it would then stand that machines and our knowledge of them is and has been undeveloped and thus ought to be examined more deeply.

VII. Conclusion

_____

Many millennia ago, a ship was wrecked off
the coast of the Greek islands of Antikythera. The
ship contained one of the most significant machines
that civilization had then seen; an analog astronomical
computer. This machine and many other classical
inventions and discoveries represent an anomaly in
the current world view of technology. We currently
do not know what it is exactly that constitutes a
machine and how machines can be delineated from
other machines. We group them into one umbrella
term of technology, and say that some require manual
power, some have a power source, some are simple,
and some are complex. But we do not have an
evidence-based means to do this.

This work has presented the idea that there
are conditions which can and should be studied that
are common to all machines, that there is a gap in the
branches of knowledge that intersects Technology,
Engineering, History, and Future(s) Studies. Machines
appear to be moving the world all around us, and yet
we have had only one word to describe each and
every one of them that has ever been invented or will
be. A brief look past the umbrella term of 'machine'

shows there are many distinguishable types, their classifications determined by evidence. Machines that rely upon different forms of energy, that have different functions or goals, and that can change or be altered differently.

Indeed, we are seeing a certain formula emerge and a unified nomenclature about which we can communicate about them. If we see a compound machine like a pair of scissors, we take the type of machine, the type of kinematic pair, the object of work, and the transformation of energy to classify it as a compound sliding-pair device. If we take this formula and apply it to a projector, we have a complex electro-mechanical apparatus. Note that ultimately, this formula varies greatly by the vast number of mechanisms and has yet to be explored more fully.

In hindsight, the way in which a study of machines may benefit us most is the simple sentiment that we will be more aware of them in our daily lives. We will be able to deconstruct the composition of each machine we see into its component parts- the parts that can be broken down into compound machines and from those into simple machines. We will be able to admire, analyze, and describe them with a common nomenclature, break down the functional fixedness of the way in which group them, and use evidence to distinguish the four types of machines. We will be able to see the degree to which each of them are involved in a task or integrated into a larger machine, and at its most basic, we will be able to understand the building blocks from which all machines originate.

A parting thought experiment is the comparison of a similar study of machines against the hypothetical of an alternative civilization. Would an alternative civilization have a study of machines or a way in which to categorize and communicate about the machines they have created in the path of their technological civilization? It may hold true that the building blocks of our machines, the six simple machines, may be identical for a non-terrestrial civilization. Thus, a basis of similarity in the language of technology between the path of all civilizations.

Final avenues of research include questions of a study of the demography of machines. To be precise, a study over time of hotspots of innovation, invention, and industry. This is pertinent to a study of machines in that the resources and individuals who create and consume machinery factor into the way in which machines change, the work that is produced, and the classification of them. Lastly, an investigation into the structure of technological revolutions[9]

[9] Cook, Scott D. N. "The Structure of Technological Revolutions and the Gutenberg Myth." *New Directions in the Philosophy of Technology* (1995): 63-83. Web.

(automation revolution, industrial revolution, etc.) may elucidate a methodology in which to study changes in craft, technique, and technology. Using the exemplar of paradigm shifts, a technical paradigm or a use of a protologism ("tech-digm") can isolate widespread shifts in the use of simple machines to compound ones, compound ones to complex ones. An explanation of this may be that the techdigm shift to complex machines in the 19th century aligns with the industrial revolution and similarly, the techdigm shift from complex ones to autonomous ones is now aligning with a current automation revolution.

In the end, we haven't had a need to classify machines because previously machines were simply slightly different parts. The different assemblies completed different functions. Now however, the parts which were simple and compound machines are contributing to mechanisms greatly beyond our understanding, most specifically, the thinking machine or artificial intelligence. With the realization of autonomous machines we see materializing what thinkers and philosophers have written about down the ages; the advent of artificial intelligence. Through a study of machines, we have the lens to classify, compare, and use evidence to understand the great spectrum of mechanisms that move the world around us.

VIII. Citations

1) Moore, Andrew (January 2011). "The hypothesis' ambassador". BioEssays. 33 (1): 1. doi:10.1002/bies.201090064.

2) Shoberg, Lore. *Machine*. New York: McGraw-Hill, 1973. Print. Encyclopedia.

3) Shoberg, Lore. *Technology*. New York: McGraw-Hill, 1973. Print. Encyclopedia.

4) Reuleaux, Franz, and Alex B. W. Kennedy. *The Kinematics of Machinery: Outlines of a Theory of Machines (1876)*. Whitefish, MT: Kessinger Publishing, 2010.

5) Arend Hintze, The 4 Types of AI, Michigan State University, 2016

6) (Kurzweil 2005, p. 260) or see Advanced Human Intelligence where he defines strong AI as "machine intelligence with the full range of human intelligence."

7) Kardashev, Nikolai (1964). "Transmission of Information by Extraterrestrial Civilizations". Soviet AstronomyWissner-Gross, Alexander D.; Freer, Cameron E. (April 19, 2013). "Causal Entropic Forces" (PDF).*Physical Review Letters*. **110**. Bibcode:2013PhRvL.110p8702W. doi:10.1103/PhysRevLett.110.168702. Retrieved 4 April 2016.

8) *Classical and Computational Solid Mechanics, YC Fung and P. Tong*. World Scientific. 2001.

9) Image from front: Antikythera Mechanism, Andrew Jerome 2016
10) Classical and Computational Solid Mechanics, *YC Fung and P. Tong*. World Scientific. 2001.
11) Cook, Scott D. N. "The Structure of Technological Revolutions and the Gutenberg Myth." *New Directions in the Philosophy of Technology* (1995): 63-83. Web.

Images

1. The Umbrella of Technology, 2016 Andrew Jerome
2. The Six Simple Machines, 2016 Andrew Jerome
3. The Mechanism of Scissors, 2016 Andrew Jerome
4. The Mechanism of a Projector, 2016 Andrew Jerome
5. The Kardashev Scale, 2016 Andrew Jerome
6. On the Classification of Machines, 2016 Andrew Jerome

Copyediting and Contributors

1. Brockton Gates, 2017
2. Louie Helm, 2017

Author contact: whochmuth@gmail.com